Feeding the Future How Entrepreneurs Can Tackle Climate Change and Ensure Food Security

J Micheal

Feeding the Future How Entrepreneurs Can Tackle Climate Change and Ensure Food Security

Copyright © 2023 by J Micheal

The first edition was published in 2023

ISBN:
Published by:
Sunshine
1663 Liberty Drive
Hyderabad, IN 47403
www.Sunshinepublishers.com

This book is self-published using on-demand printing and publishing, which allows it to be printed and distributed globally.

TABLE OC CONTENT

Chapter 1: Understanding the Climate Change-Food Security Nexus

The Impact of Climate Change on Food Systems

Climate change is one of the greatest challenges of our time, and its impact on global food systems cannot be overstated. As entrepreneurs, understanding the intricate relationship between climate change and food security is crucial for developing innovative solutions that can help us navigate these uncertain times.

One of the primary ways climate change affects food systems is through extreme weather events. Rising temperatures, changing precipitation patterns, and an increase in the frequency and intensity of storms are disrupting agricultural practices around the world. Droughts and floods are becoming more common, leading to crop failures, reduced yields, and loss of livelihoods for farmers. As entrepreneurs, it is essential to recognize the vulnerabilities created by these climate-related disasters and work towards building resilient and adaptive food systems.

Furthermore, climate change is altering the distribution and availability of natural resources necessary for food production. Changes in water availability, soil quality, and pollinator populations are affecting crop productivity and diversity. Entrepreneurs can play a significant role in developing sustainable farming practices, such as precision agriculture and regenerative farming techniques, which optimize resource use and minimize environmental impact.

Another critical aspect to consider is the impact of climate change on food security and nutrition. Climate-related factors can disrupt supply chains, leading to increased food prices and reduced access to nutritious food, particularly for vulnerable populations. Entrepreneurs can contribute by developing innovative solutions to enhance food storage and preservation, improve food distribution networks, and promote sustainable and locally sourced food options.

The role of entrepreneurs in addressing climate change and ensuring food security goes beyond technological innovations. It also involves fostering partnerships, knowledge sharing, and policy advocacy. Collaboration between entrepreneurs, governments, and civil society organizations is crucial for scaling up solutions and creating an enabling environment for sustainable food systems.

In conclusion, climate change poses significant challenges to global food systems, impacting agricultural practices, resource availability, and food security. As entrepreneurs, it is essential to recognize these challenges and harness our creativity and innovative spirit to develop sustainable solutions. By embracing resilient farming practices, promoting sustainable food production and distribution, and advocating for supportive policies, we can contribute to feeding the future and building a more climate-resilient world. Let us seize the opportunities presented by these challenges to create a future where food security and climate action go hand in hand.

The Relationship between Climate Change and Food Security

In recent years, there has been growing concern about the impact of climate change on food security. Climate change refers to long-term alterations in temperature, precipitation, wind patterns, and other aspects of the Earth's climate system. These changes have far-reaching effects on the availability, accessibility, and quality of food, thereby posing significant challenges to global food security. As entrepreneurs in the field of climate change, it is crucial to understand the intricate relationship between climate change and food security and explore innovative solutions to address these challenges.

One of the primary ways in which climate change affects food security is through its impact on agricultural productivity. Rising temperatures, changing rainfall patterns, and extreme weather events such as droughts, floods, and storms can all result in reduced crop yields and livestock productivity. This not only threatens the livelihoods of farmers but also disrupts food supply chains, leading to increased food prices and reduced availability of nutritious food for vulnerable populations.

Furthermore, climate change also affects the nutritional content of food. Higher levels of carbon dioxide in the atmosphere, combined with temperature changes, can lead to reduced levels of essential nutrients in crops, such as iron, zinc, and protein. This can have severe consequences for human health, particularly in regions already experiencing high rates of malnutrition.

The relationship between climate change and food security is complex and multifaceted. However, entrepreneurs have a unique opportunity to contribute to the solution. By developing innovative technologies and business models, entrepreneurs can help mitigate climate change impacts on agriculture and enhance food security.

For instance, entrepreneurs can invest in climate-smart agricultural practices that promote sustainable land management, efficient water use, and the conservation of biodiversity. These practices not only help reduce greenhouse gas emissions but also enhance the resilience of agricultural systems to climate change. Additionally, entrepreneurs can leverage advancements in technology, such as precision agriculture, remote sensing, and data analytics, to optimize resource use, improve crop yields, and reduce post-harvest losses.

Collaboration and partnerships between entrepreneurs, governments, research institutions, and local communities are also vital for addressing the complex challenges at the intersection of climate change and food security. By working together, we can develop scalable and sustainable solutions that ensure food security for all, even in the face of a changing climate.

In conclusion, the relationship between climate change and food security is a pressing issue that requires urgent attention. As entrepreneurs in the niche of climate change, it is our responsibility to understand these challenges and find innovative solutions. By promoting sustainable agricultural practices, leveraging technology, and fostering collaboration, we can contribute to a future where food security is not compromised by climate change. Together, we can feed the future.

Current Challenges and Future Projections

In today's world, climate change poses significant challenges to global food security. The impact of rising temperatures, changing rainfall patterns, and extreme weather events have already started to affect agricultural production and threaten the availability of food for the growing global population. As entrepreneurs, it is crucial for us to understand these challenges and explore innovative solutions to ensure a sustainable future for food production and climate resilience.

One of the primary challenges we face is the increasing frequency and intensity of extreme weather events. Droughts, floods, and storms can devastate crops, leading to reduced yields and increased food prices. To combat this, entrepreneurs need to invest in climate-smart technologies and practices that can enhance resilience in agriculture. This includes precision farming techniques, such as remote sensing and data analytics that enable farmers to optimize resource management and reduce environmental impact.

Another pressing challenge is the loss of biodiversity and degradation of ecosystems. Climate change disrupts natural habitats, leading to the loss of plant and animal species essential for maintaining a healthy ecosystem. Entrepreneurs can play a crucial role in developing sustainable agricultural practices that promote biodiversity, such as agroforestry systems and organic farming methods. By restoring and protecting ecosystems, we can enhance the resilience of food systems and ensure long-term food security.

Furthermore, the increasing demand for food due to population growth exacerbates the challenges posed by climate change. By 2050, the global population is projected to reach 9.7 billion, requiring a significant increase in food production. However, traditional agricultural practices are often resource-intensive and contribute to greenhouse gas emissions. Entrepreneurs can address this challenge by investing in alternative food production systems, such as vertical farming and aquaponics, which require less land, water, and energy. Additionally, entrepreneurs can explore plant-based alternatives to animal protein, reducing the environmental impact of livestock farming.

Looking ahead, the future projections for food security and climate change are intertwined. Without proactive measures, climate change will continue to pose risks to our food systems, exacerbating hunger and poverty. However, by embracing innovative technologies, sustainable practices, and collaboration among stakeholders, entrepreneurs can drive the transformation needed to ensure a resilient and secure food future.

In conclusion, the current challenges posed by climate change to food security require urgent attention and innovative solutions from entrepreneurs. By investing in climate-smart technologies, promoting biodiversity, and exploring alternative food production systems, we can tackle climate change and ensure a sustainable future for global food security. It is our responsibility as entrepreneurs to rise to this challenge and create a positive impact on the world by feeding the future.

14

Chapter 2: The Role of Entrepreneurs in Addressing Climate Change and Food Security

The Power of Entrepreneurship in Driving Change

Introduction:

In the face of climate change and the growing need for food security, entrepreneurs have emerged as key players in driving positive change. Their innovative ideas, risk-taking mindset, and ability to mobilize resources have positioned them to tackle the complex challenges we face today. This subchapter explores the transformative power of entrepreneurship in addressing climate change and ensuring food security, highlighting the crucial role entrepreneurs play in shaping a sustainable future.

Entrepreneurship as a Catalyst for Innovation: Entrepreneurs are known for their ability to identify gaps in the market and develop innovative solutions. In the context of climate change and food security, entrepreneurs are creating groundbreaking technologies and disruptive business models that promote sustainable practices. From vertical farming and precision agriculture to alternative protein sources and food waste reduction initiatives, entrepreneurs are at the forefront of revolutionizing the way we produce, distribute, and consume food.

Entrepreneurs as Agents of Change: Entrepreneurs possess a unique set of skills and characteristics that enable them to be effective agents of change. Their tenacity, resilience, and ability to take risks allow them to challenge the status quo and drive transformation. By breaking down barriers and pushing boundaries, entrepreneurs create new opportunities for sustainable practices, not only within their own ventures but also across industries. They inspire others to think differently and contribute to a collective effort in combating climate change and ensuring food security.

Collaboration and Partnerships: Entrepreneurs understand that collaboration is key to achieving meaningful impact. They actively seek partnerships with governments, NGOs, research institutions, and other stakeholders to amplify their efforts. By leveraging diverse expertise and resources, entrepreneurs can scale their solutions and address complex challenges more effectively. Collaboration also fosters knowledge sharing, creating a collective intelligence that can drive systemic change.

Policy Advocacy and Influence: Entrepreneurs have a unique ability to influence policymakers and shape regulations. Through successful ventures and innovative solutions, they demonstrate the feasibility and economic viability of sustainable practices. By effectively communicating the benefits of their initiatives, entrepreneurs can sway public opinion and inspire policymakers to enact supportive policies. Additionally, entrepreneurs can form alliances and associations to collectively advocate for policies that promote sustainable practices and address climate change and food security.

Conclusion:

The power of entrepreneurship in driving change cannot be underestimated. Entrepreneurs are at the forefront of the fight against climate change and the quest for food security. Through their innovative ideas, resilience, collaboration, and policy advocacy, they are shaping a sustainable future for generations to come. As entrepreneurs, it is our responsibility to recognize the immense power we possess and use it to create positive change that benefits not only our businesses but also the planet and its inhabitants. Together, we can tackle climate change and ensure food security, paving the way for a better tomorrow.

Innovative Solutions for Sustainable Agriculture

Introduction:

In recent years, the world has witnessed the effects of climate change on our planet, including its significant impact on agriculture and food security. As entrepreneurs, we have a unique opportunity and responsibility to develop innovative solutions that can tackle climate change and ensure sustainable agriculture for future generations. This subchapter explores some of the most promising approaches and technologies that can revolutionize the way we produce and consume food.

1. Precision Agriculture:

Precision agriculture employs advanced technologies such as remote sensing, drones, and data analytics to optimize farming practices. By collecting real-time data on soil conditions, water availability, and crop health, entrepreneurs can develop smart farming solutions that minimize resource wastage, reduce greenhouse gas emissions, and enhance overall productivity.

2. Vertical Farming:

With a rapidly growing global population, traditional farming methods are becoming increasingly unsustainable. Vertical farming offers an innovative solution by utilizing vertical space to grow crops indoors, using hydroponics or aeroponics systems. This approach dramatically reduces water consumption, eliminates the need for harmful pesticides, and allows for year-round cultivation, irrespective of weather conditions.

3.   Agroforestry:
Agroforestry combines the principles of agriculture and forestry, integrating trees with crops and livestock. By diversifying agricultural landscapes, entrepreneurs can enhance soil fertility, reduce erosion, sequester carbon, and provide additional sources of income through timber or non-timber forest products. Agroforestry systems also promote biodiversity and enhance resilience against climate change impacts.

4.   Plant-Based   Alternatives:
The global demand for meat and dairy products is putting tremendous pressure on the environment. Entrepreneurs can develop plant-based alternatives that replicate the taste and texture of animal products, providing consumers with sustainable and healthy options. These innovations not only reduce greenhouse gas emissions and water usage but also address ethical concerns related to animal welfare.

5.   Food   Waste   Reduction:
Food waste is a significant contributor to greenhouse gas emissions and resource depletion. Entrepreneurs can tackle this issue by developing technologies to monitor and reduce food waste throughout the supply chain. From smart packaging solutions to innovative food recycling methods, these approaches can minimize waste, conserve resources, and alleviate the pressure on agricultural systems.

Conclusion:

As entrepreneurs in the face of climate change, we have the power to drive meaningful change in the agricultural sector. By embracing innovative solutions such as precision agriculture, vertical farming, agroforestry, plant-based alternatives, and food waste reduction, we can ensure a sustainable future for agriculture while mitigating climate change. Let us seize this opportunity to feed the future and leave a positive impact on both our planet and the communities that depend on it.

Entrepreneurial Opportunities in Climate-Smart Food Systems

Introduction:

In recent years, climate change and its impact on food security have become pressing global concerns. As entrepreneurs, there is a unique opportunity to address these challenges head-on and contribute to building climate-smart food systems. This subchapter explores the various entrepreneurial opportunities that exist in this niche, offering insights and guidance for entrepreneurs looking to make a positive impact on both climate change and food security.

1. Sustainable Agriculture:

One of the key areas where entrepreneurs can make a significant difference is in promoting sustainable agriculture practices. This involves adopting climate-smart farming techniques such as precision agriculture, agroforestry, and organic farming. Entrepreneurs can develop innovative solutions like smart irrigation systems, vertical farming technologies, and soil health monitoring devices. These solutions not only reduce greenhouse gas emissions but also improve agricultural productivity and resilience.

2. Food Waste Reduction:

Another critical area for entrepreneurial intervention is food waste reduction. Approximately one-third of global food production is wasted, contributing to both climate change and food insecurity. Entrepreneurs can explore opportunities in food recovery and redistribution, developing technologies and platforms that connect surplus food with those in need. Additionally, innovative packaging and storage solutions can help extend the shelf life of perishable foods, reducing waste along the supply chain.

3. Renewable Energy Integration:

Entrepreneurs can play a crucial role in integrating renewable energy sources into the food system. This includes developing and deploying solar-powered cold storage units, biomass energy systems, and clean cooking technologies. By shifting from fossil fuel-based energy to renewable sources, entrepreneurs can significantly reduce the carbon footprint of the food system, while also improving energy efficiency and reducing costs.

4. Supply Chain Optimization:

Optimizing the food supply chain is another area where entrepreneurial opportunities lie. This involves leveraging technology to track and trace food products, ensuring transparency and accountability throughout the supply chain. Entrepreneurs can also develop innovative logistics solutions to reduce food spoilage during transportation and storage, thereby minimizing greenhouse gas emissions and improving overall efficiency.

Conclusion:

Entrepreneurs have a unique opportunity to drive transformative change in climate-smart food systems. By embracing sustainable agriculture, reducing food waste, integrating renewable energy, and optimizing the food supply chain, entrepreneurs can contribute to both climate change mitigation and food security. This subchapter has provided a glimpse of the vast entrepreneurial opportunities available within this niche. It is time for entrepreneurs to rise to the challenge and take action, ensuring a sustainable and secure future for all.

Chapter 3: Building Resilient Food Systems: Strategies for Entrepreneurs

Sustainable Farming Practices and Technologies

In recent years, the world has witnessed the devastating impact of climate change on our food systems. Entrepreneurs have a significant role to play in addressing this global challenge and ensuring food security for future generations. This subchapter explores the various sustainable farming practices and technologies that can help entrepreneurs tackle climate change and foster food security.

Sustainable farming practices prioritize environmental stewardship, social responsibility, and economic viability. By implementing these practices, entrepreneurs can minimize the carbon footprint of agriculture, conserve natural resources, and ensure the long-term productivity of their farms. One such practice is organic farming, which eliminates the use of synthetic pesticides and fertilizers, promoting soil health and biodiversity.

Another crucial aspect is regenerative agriculture, which aims to restore degraded soil and enhance its capacity to sequester carbon. Entrepreneurs can adopt techniques such as cover cropping, crop rotation, and conservation tillage to improve soil fertility, reduce erosion, and mitigate climate change. Agroforestry, the integration of trees and crops, provides multiple benefits, including carbon sequestration, biodiversity preservation, and improved water management.

Technological advancements have also revolutionized sustainable farming. Precision agriculture utilizes sensors, drones, and data analytics to optimize resource use, minimize waste, and increase crop yields. By precisely monitoring soil moisture, nutrient levels, and pest infestations, entrepreneurs can make informed decisions, reducing the need for excessive inputs and enhancing efficiency.

Additionally, vertical farming and hydroponics offer innovative solutions for urban agriculture. By growing crops in vertically stacked layers or nutrient-rich water solutions, entrepreneurs can maximize land use, reduce water consumption, and minimize transportation costs. These technologies also allow for year-round production, ensuring a consistent food supply regardless of seasonal fluctuations.

Furthermore, entrepreneurs can explore the potential of emerging technologies like artificial intelligence and block chain in sustainable farming. AI can support decision-making processes, optimize resource allocation, and predict crop diseases. Block chain technology can enhance transparency and traceability in the food supply chain, ensuring fair practices and reducing food waste.

In conclusion, sustainable farming practices and technologies are crucial tools for entrepreneurs to tackle climate change and ensure food security. By adopting organic farming, and regenerative agriculture, and embracing technological advancements, entrepreneurs can contribute to mitigating the impacts of climate change, protecting natural resources, and creating a sustainable future for agriculture. As entrepreneurs, it is our responsibility to embrace innovation and foster partnerships to secure a resilient and nourished future for all.

Climate-Resilient Crop Selection and Diversification

In the face of climate change, the agricultural sector must adapt to ensure food security for the growing global population. Entrepreneurs play a crucial role in this process by innovating and implementing sustainable solutions. One such solution is climate-resilient crop selection and diversification.

Climate change presents numerous challenges to crop production, including extreme weather events, changing rainfall patterns, and the spread of pests and diseases. To mitigate these risks and ensure food production, entrepreneurs need to focus on selecting and diversifying crops that are better adapted to the changing climate.

The first step in climate-resilient crop selection is to identify which crops are more likely to thrive under specific climate conditions. Entrepreneurs can utilize advanced technologies, such as satellite imagery and climate modelling, to assess the suitability of different crops in a particular region. This data-driven approach allows entrepreneurs to make informed decisions about which crops to grow, optimizing resource allocation and reducing the risk of crop failure.

Diversification is another key strategy in climate-resilient crop selection. By growing a variety of crops with different characteristics, entrepreneurs can hedge against climate-related risks. For example, some crops may be more tolerant to drought, while others may have higher resistance to pests and diseases. By diversifying their crop portfolio, entrepreneurs can ensure a more stable and reliable food supply, even in the face of changing environmental conditions.

Entrepreneurs can also explore the potential of traditional and indigenous crops that have shown resilience to local climates. These crops, often overlooked by conventional agriculture, may offer valuable alternatives that are better adapted to future climate scenarios. By promoting the cultivation and consumption of these crops, entrepreneurs can not only enhance food security but also preserve traditional knowledge and support local communities.

Furthermore, entrepreneurs can leverage technological advancements to optimize crop production and minimize the environmental impact. Precision agriculture techniques, such as smart irrigation systems and remote sensing, enable efficient resource management and reduce water and fertilizer use. By adopting these technologies, entrepreneurs can enhance the resilience of their crops while minimizing their carbon footprint.

In conclusion, climate-resilient crop selection and diversification are crucial strategies for entrepreneurs to tackle climate change and ensure food security. By leveraging advanced technologies, diversifying crop portfolios, and exploring traditional and indigenous crops, entrepreneurs can adapt to the changing climate and contribute to a more sustainable and resilient agricultural sector.

Efficient Water Management and Irrigation Techniques

In the face of climate change, one of the most pressing challenges entrepreneurs in the agricultural sector face is ensuring food security while conserving our precious water resources. With changing weather patterns and increased water scarcity, adopting efficient water management and irrigation techniques has become crucial for sustainable farming practices. This subchapter aims to provide entrepreneurs in the field with an overview of innovative approaches to address this issue.

1.   Understanding   Water   Management: To effectively manage water resources, entrepreneurs must first understand the water cycle and its implications for their farming operations. This includes analysing rainfall patterns, evaporation rates, and water availability in their region. Armed with this knowledge, entrepreneurs can develop tailored strategies to optimize water usage and reduce wastage.

2.   Implementing   Precision   Irrigation   Techniques: Precision irrigation techniques, such as drip irrigation and micro-sprinklers, have proven to be highly efficient in delivering water directly to plant roots. By adopting these techniques, entrepreneurs can significantly reduce water loss due to evaporation, runoff, and deep percolation. Moreover, precision irrigation can enhance crop yields and quality, leading to improved profitability.

3.   Embracing   Smart   Irrigation   Technology: Harnessing the power of technology, entrepreneurs can now

deploy smart irrigation systems that use sensors and data analytics to optimize water usage. These systems can monitor soil moisture levels, weather conditions, and crop water requirements in real time, allowing for precise irrigation scheduling. By automating irrigation, entrepreneurs can further reduce water waste and ensure that plants receive just the right amount of water when they need it.

4. Employing Water Recycling and Reuse: Entrepreneurs can explore innovative techniques such as rainwater harvesting and water recycling to minimize their reliance on freshwater sources. Rainwater can be collected and stored for irrigation purposes, while treated wastewater can be reused for non-potable needs. By implementing these strategies, entrepreneurs can minimize the strain on local water supplies and contribute to sustainable water management.

5. Educating Farmers and Stakeholders: Entrepreneurs play a vital role in promoting the adoption of efficient water management practices among farmers and other stakeholders. By organizing training programs, workshops, and awareness campaigns, entrepreneurs can empower individuals with knowledge about water conservation, irrigation techniques, and the economic benefits of sustainable farming. Additionally, entrepreneurs can collaborate with governments, NGOs, and research institutions to advocate for policies and incentives that encourage water-efficient practices.

In conclusion, efficient water management and irrigation techniques are essential for entrepreneurs to tackle climate change and ensure food security. By embracing precision irrigation, smart technology, water recycling, and educating stakeholders, entrepreneurs can pave the way for a sustainable and resilient agricultural sector that thrives in a changing climate. Let us seize the opportunity to be at the forefront of this transformation and contribute to a more food-secure future.

Chapter 4: Scaling Up Sustainable Food Production: Entrepreneurial Approaches

Investing in Agri-tech Start-ups

As entrepreneurs, it is crucial to recognize the pressing challenges posed by climate change and food security. The future of our planet and the well-being of humanity depend on finding innovative solutions to these issues. Agri-tech startups have emerged as a promising avenue for addressing these concerns, offering the potential to revolutionize the agricultural industry while mitigating the impact of climate change. In this subchapter, we will explore the reasons why investing in agri-tech startups is not only a financially sound decision but also a powerful way to tackle climate change and ensure food security.

Agri-tech startups leverage cutting-edge technologies, such as artificial intelligence, precision farming, robotics, and data analytics, to optimize agricultural practices. By harnessing the power of these technologies, these startups are capable of increasing crop yields, reducing resource consumption, and minimizing environmental impact. This approach aligns perfectly with the needs of our rapidly changing climate and the increasing demand for food worldwide.

Investing in agri-tech startups presents an incredible opportunity to drive positive change while generating significant financial returns. The global agri-tech market is experiencing rapid growth, with a projected value of over $22 billion by 2025. Entrepreneurs who recognize this potential and invest early in promising agri-tech startups stand to gain not only financial rewards but also the satisfaction of being part of a transformative movement.

Moreover, investing in agri-tech startups allows entrepreneurs to actively contribute to combating climate change. Agriculture is a major contributor to greenhouse gas emissions, deforestation, and water pollution. By supporting startups that develop sustainable farming practices, resource-efficient technologies, and climate-resilient crops, entrepreneurs can play a pivotal role in reducing the environmental footprint of the agricultural sector.

Furthermore, investing in agri-tech startups fosters food security by increasing agricultural productivity and resilience. As the world's population continues to grow, ensuring a stable and abundant food supply becomes increasingly challenging. Agri-tech startups offer innovative solutions to address these concerns, such as vertical farming, hydroponics, and smart irrigation systems. By investing in these startups, entrepreneurs can help create a more sustainable and secure food system for future generations.

In conclusion, investing in agri-tech startups provides entrepreneurs with an incredible opportunity to tackle climate change and ensure food security while generating significant financial returns. By supporting these startups, entrepreneurs can actively contribute to the development of sustainable farming practices, resource-efficient technologies, and climate-resilient crops. The agri-tech sector is poised for exponential growth, making early investments highly lucrative. Together, as entrepreneurs, we have the power to shape a sustainable future, and investing in agri-tech startups is a powerful way to do so.

34

Vertical Farming and Urban Agriculture

Vertical Farming and Urban Agriculture: Revolutionizing Food Production for a Sustainable Future

In recent years, the world has witnessed the rapid growth of urbanization, with more people than ever before living in cities. This unprecedented shift has put tremendous pressure on traditional farming practices, leading to concerns about food security and the impact of agriculture on climate change. However, entrepreneurs have stepped up to address these challenges through innovative solutions such as vertical farming and urban agriculture.

Vertical farming refers to the practice of growing crops in vertically stacked layers, often in a controlled environment such as a greenhouse or an indoor facility. By utilizing advanced technologies like hydroponics and aeroponics, vertical farms can optimize resource utilization, reduce water consumption, and minimize the use of pesticides and fertilizers. This approach not only maximizes crop yield but also ensures year-round production, mitigating the impact of seasonal variations.

One of the key advantages of vertical farming is its ability to bring food production closer to urban areas, reducing transportation distances and carbon emissions associated with long-distance food supply chains. Urban agriculture, on the other hand, focuses on integrating food production within cities themselves. This can include rooftop gardens, community gardens, and even repurposed buildings transformed into productive farms. By utilizing underutilized spaces, urban agriculture enhances access to fresh produce and reduces the strain on rural agricultural areas.

Entrepreneurs in the field of vertical farming and urban agriculture have been instrumental in transforming the food production landscape. By leveraging technology, data analytics, and automation, they have optimized growing conditions, minimized waste, and improved efficiency. These advancements have not only made urban agriculture economically viable but also scalable, providing an opportunity to meet the increasing demand for food sustainably.

Furthermore, vertical farming and urban agriculture have the potential to address other critical challenges linked to climate change. By sequestering carbon dioxide, these practices contribute to reducing greenhouse gas emissions. Additionally, they can help in managing water resources more efficiently, as vertical farming uses up to 90% less water compared to traditional agriculture.

As entrepreneurs, the opportunities in vertical farming and urban agriculture are immense. By investing in this sector, you have the potential to make a significant impact on food security, climate change mitigation, and sustainable development. From developing innovative farming systems to creating new market opportunities for locally grown produce, the possibilities for entrepreneurship in this field are endless.

Feeding the Future is a comprehensive guide that explores the potential of vertical farming and urban agriculture while providing valuable insights and practical advice for entrepreneurs looking to enter this exciting industry. By embracing these technologies and practices, entrepreneurs can not only contribute to a more sustainable food system but also create profitable business ventures that address the urgent challenges of climate change and ensure a secure food future for generations to come.

Promoting Local Food Systems and Short Supply Chains

In today's world, where climate change poses a significant threat to our planet and food security is becoming an increasingly pressing issue, entrepreneurs have a unique opportunity to make a positive impact. One effective approach to tackle these challenges is by promoting local food systems and short supply chains. By doing so, entrepreneurs can not only contribute to mitigating climate change but also ensure food security for communities around the world.

Local food systems are characterized by the production, distribution, and consumption of food that is locally sourced, often within a defined geographical area. Short supply chains, on the other hand, involve the direct exchange of goods between producers and consumers, minimizing the number of intermediaries. These approaches offer several benefits that make them attractive for entrepreneurs looking to address climate change and food security.

Firstly, promoting local food systems and short supply chains helps reduce greenhouse gas emissions associated with long-distance transportation of food. By sourcing food locally, entrepreneurs can minimize the carbon footprint of their operations, contributing to the fight against climate change. Additionally, shorter supply chains mean less food waste, as products spend less time in transit and have a higher chance of being consumed before spoiling.

Furthermore, local food systems and short supply chains offer economic advantages for entrepreneurs. By connecting local producers directly with consumers, entrepreneurs can foster a sense of community and support local economies. Small-scale farmers and producers can gain fairer prices for their products, leading to increased profitability and improved livelihoods. Entrepreneurs can also tap into the growing consumer demand for fresh, locally sourced food, creating new business opportunities and building customer loyalty.

In addition to the environmental and economic benefits, promoting local food systems and short supply chains enhances food security. By reducing dependence on global food supply chains, entrepreneurs can help communities become more self-reliant in terms of food production. This resilience is crucial in the face of climate-related disruptions and other potential threats to global food systems.

To promote local food systems and short supply chains, entrepreneurs can explore various strategies. These may include establishing farmers' markets, community-supported agriculture programs, or farm-to-table restaurants. Collaborating with local farmers and producers, supporting urban agriculture initiatives, and investing in innovative technologies that enhance local food production are also effective approaches.

In conclusion, promoting local food systems and short supply chains presents a significant opportunity for entrepreneurs to address climate change and ensure food security. By embracing these approaches, entrepreneurs can reduce greenhouse gas emissions, support local economies, and enhance community resilience. Through their innovative and sustainable efforts, entrepreneurs can play a crucial role in feeding the future while safeguarding our planet.

Chapter 5: Collaboration and Partnerships for Food Security

Public-Private Partnerships in Agricultural Innovation

In recent years, the world has witnessed the increasing impact of climate change on our planet's ecosystems, including the agricultural sector. Entrepreneurs, who are often at the forefront of innovation and problem-solving, have a unique role to play in tackling climate change and ensuring food security. One powerful approach that entrepreneurs can leverage is through public-private partnerships (PPPs) in agricultural innovation.

PPPs are collaborative initiatives between government entities and private sector organizations that aim to address complex societal challenges. In the context of climate change and food security, PPPs can bring together the expertise, resources, and networks of both sectors to drive innovation, foster sustainable practices, and create lasting impact.

The agricultural sector is particularly vulnerable to the effects of climate change, such as extreme weather events, shifting growing seasons, and increased pest and disease pressure. Entrepreneurs have a crucial role in developing innovative solutions that can help farmers adapt and mitigate these challenges. However, they often face barriers such as limited access to funding, lack of infrastructure, and limited market opportunities. This is where PPPs can play a pivotal role in supporting and scaling up entrepreneurial efforts.

By partnering with government agencies, entrepreneurs can gain access to public funding, research facilities, and regulatory support. This collaboration can enable entrepreneurs to develop and test new technologies, such as precision agriculture tools, climate-smart crop varieties, and efficient irrigation systems. PPPs can also facilitate the transfer of knowledge and best practices between researchers, policymakers, and entrepreneurs, leading to more effective and targeted solutions.

Furthermore, PPPs can help entrepreneurs navigate market challenges by facilitating partnerships with private sector actors, including retailers, distributors, and investors. These partnerships can create market demand for innovative products and technologies, ensuring their adoption and scalability. By working together, entrepreneurs and private sector partners can build sustainable value chains that promote climate resilience, reduce waste, and improve access to nutritious food.

To fully harness the potential of PPPs in agricultural innovation, entrepreneurs must actively engage with government agencies, research institutions, and other stakeholders. They should seek out collaboration opportunities, participate in innovation competitions and funding programs, and advocate for supportive policies that foster entrepreneurship and sustainability.

In conclusion, entrepreneurs have a critical role to play in addressing climate change and ensuring food security. Public-private partnerships in agricultural innovation can provide entrepreneurs with the necessary support, resources, and networks to develop and scale up their solutions. By leveraging these partnerships, entrepreneurs can drive sustainable agricultural practices, promote resilience in the face of climate change, and contribute to a more food-secure future for all.

Engaging Smallholder Farmers in Climate Change Adaptation

In recent years, the issue of climate change has gained significant attention due to its wide-ranging impacts on our environment and society. One of the most vulnerable groups affected by climate change is smallholder farmers, who rely heavily on agriculture for their livelihoods. As entrepreneurs in the field of climate change, it is essential to understand the unique challenges faced by these farmers and devise innovative solutions to help them adapt to the changing climate.

Smallholder farmers play a crucial role in ensuring food security, especially in developing countries. However, their agricultural practices are often traditional and not equipped to withstand the adverse effects of climate change. Rising temperatures, erratic rainfall patterns, and increased pest and disease outbreaks are just a few examples of the challenges they face. As entrepreneurs, it is our responsibility to engage with smallholder farmers and empower them to adapt their farming practices in the face of these challenges.

Engagement begins with education and awareness. Many smallholder farmers are unaware of the long-term implications of climate change on their agricultural activities. By providing them with knowledge and information about climate change, its impacts, and potential adaptation strategies, we can help them make informed decisions about their farming practices. This can be done through workshops, training programs, and the dissemination of relevant information via various media channels.

Furthermore, entrepreneurs can play a crucial role in facilitating access to climate-smart technologies and practices for smallholder farmers. These technologies include drought-resistant crops, efficient irrigation systems, and improved soil management techniques. By partnering with technology providers, governments, and NGOs, entrepreneurs can ensure that these solutions reach the hands of those who need them the most. This not only helps farmers adapt to climate change but also enhances their productivity and income, ultimately contributing to food security.

Collaboration is key in engaging smallholder farmers in climate change adaptation. Entrepreneurs can foster partnerships with local communities, agricultural cooperatives, and other stakeholders to develop holistic and context-specific solutions. By involving farmers in the decision-making process and understanding their needs and challenges, entrepreneurs can create innovative business models that are mutually beneficial and sustainable in the long run.

In conclusion, engaging smallholder farmers in climate change adaptation is not only a moral imperative but also presents a significant business opportunity for entrepreneurs. By providing education, facilitating access to climate-smart technologies, and fostering collaboration, entrepreneurs can empower smallholder farmers to adapt to climate change, ensure food security, and create a more sustainable future for all.

Collaborative Initiatives for Knowledge Sharing and Capacity Building

In the face of climate change and the growing threat to global food security, entrepreneurs have a crucial role to play. The challenges ahead require innovative solutions and collaborative efforts to mitigate the impact of climate change on our food systems. This subchapter explores various collaborative initiatives for knowledge sharing and capacity building that can empower entrepreneurs to tackle climate change and ensure food security.

Knowledge sharing is a fundamental aspect of addressing climate change and food security. Entrepreneurs need access to up-to-date information, best practices, and scientific research to develop sustainable and resilient solutions. Collaborative platforms, such as online communities, forums, and conferences, play a vital role in connecting entrepreneurs, researchers, and policymakers. These platforms facilitate the exchange of ideas, experiences, and expertise, fostering innovation and enabling entrepreneurs to stay informed about the latest advancements in climate-smart agriculture and sustainable food production.

Capacity building is another crucial element for entrepreneurs seeking to address climate change and food security. Building the skills and capabilities necessary to navigate complex challenges is essential. Collaborative initiatives can provide training programs, mentorship, and networking opportunities to enhance entrepreneurs' capacity to develop and implement sustainable solutions. By connecting entrepreneurs with experienced professionals and mentors, these initiatives offer guidance, support, and valuable insights.

Public-private partnerships are a powerful collaborative approach to addressing climate change and food security. Entrepreneurs can partner with governments, NGOs, research institutions, and other stakeholders to leverage resources, expertise, and networks. These partnerships can lead to the development of innovative policies, funding opportunities, and access to diverse markets. By working together, entrepreneurs can amplify their impact, create synergies, and foster sustainable change at a larger scale.

International collaborations are also vital for entrepreneurs tackling climate change and food security. Sharing knowledge and expertise across borders can accelerate the development and adoption of sustainable practices. Collaboration between entrepreneurs from different regions allows for the exchange of diverse perspectives, cultural insights, and unique solutions. International networks and partnerships can facilitate technology transfer, market access, and capacity building, enabling entrepreneurs to make a global impact.

In conclusion, collaborative initiatives for knowledge sharing and capacity building are essential for entrepreneurs addressing climate change and ensuring food security. By actively participating in collaborative platforms, engaging in public-private partnerships, and embracing international collaborations, entrepreneurs can leverage collective knowledge and resources to develop innovative, sustainable, and scalable solutions. Together, entrepreneurs can play a pivotal role in building a resilient and food-secure future in the face of climate change.

49

Chapter 6: Financing Sustainable Food Ventures

Accessing Climate Finance for Agri-businesses

Climate change presents a significant challenge to the agri-business sector, as it threatens food security and the stability of agricultural production. However, entrepreneurs have the power to tackle this issue head-on and contribute to a sustainable future. One avenue for achieving this is by accessing climate finance, which provides the necessary funding for agri-businesses to implement climate-smart practices and technologies.

Climate finance refers to financial resources that are dedicated to climate change mitigation, adaptation, and resilience efforts. It aims to support projects and initiatives that reduce greenhouse gas emissions, enhance climate resilience, and promote sustainable development. For agri-businesses, accessing climate finance can open up a world of opportunities to transform their operations and become more climate-resilient.

To tap into climate finance, entrepreneurs in the agri-business sector need to be proactive and well-prepared. Here are some essential steps to consider:

1. Understand the Climate Finance Landscape: Familiarize yourself with the various sources of climate finance, such as international funds, national climate finance mechanisms, and private sector initiatives. Research the eligibility criteria and funding priorities of these sources to identify the most suitable ones for your agri-business.

2. Develop a Climate-Smart Business Plan: Craft a comprehensive business plan that outlines how your agri-business will address climate change challenges and contribute to sustainable development. Highlight the specific climate-smart practices and technologies you intend to implement, along with their expected environmental and financial benefits.

3. Build Partnerships and Collaborations: Climate finance often requires collaboration with other stakeholders, such as government agencies, NGOs, and financial institutions. Establish partnerships and networks with these organizations to enhance your chances of accessing climate finance and benefit from their technical expertise.

4. Gather Relevant Data and Information: Climate finance providers require robust data and information to assess the feasibility and impact of your proposed projects. Collect and organize data related to your agri-business operations, such as greenhouse gas emissions, water usage, and energy consumption. This information will help demonstrate the potential environmental benefits of your initiatives.

5. Strengthen Financial Management and Reporting: Effective financial management and reporting are crucial for securing climate finance. Implement robust financial systems and reporting mechanisms that track and document the use of climate finance funds. This will enhance accountability and transparency, increasing the likelihood of future funding.

6. Seek Professional Assistance: Accessing climate finance can be complex and time-consuming. Consider seeking guidance from climate finance experts or consultants who specialize in assisting agri-businesses. They can provide valuable insights, assist in preparing funding proposals, and navigate the intricacies of the application process.

By accessing climate finance, agri-business entrepreneurs can accelerate the adoption of climate-smart practices, enhance their resilience to climate change impacts, and contribute to global efforts in mitigating climate change. The funding obtained through climate finance can support investments in renewable energy technologies, sustainable irrigation systems, precision agriculture, and climate-resilient crop varieties. Ultimately, this will ensure food security, promote sustainable development, and create a brighter future for both the agri-business sector and the planet.

Impact Investing and Social Entrepreneurship

In recent years, there has been a growing recognition of the need for entrepreneurs to address climate change and ensure food security. As traditional approaches have proven insufficient in tackling these global challenges, a new paradigm has emerged: impact investing and social entrepreneurship. This subchapter will dive into the concept of impact investing and its role in driving positive change in the face of climate change, while also exploring the domain of social entrepreneurship and its potential to create sustainable solutions for food security.

Impact investing is an investment strategy that seeks both financial returns and positive social or environmental impact. Unlike traditional investing, which primarily focuses on financial gains, impact investing aims to generate measurable, beneficial outcomes for society and the planet. By directing capital towards businesses and initiatives that align with the goals of climate change mitigation and food security, impact investors play a vital role in driving innovation and change.

Social entrepreneurship, on the other hand, refers to the practice of using entrepreneurial principles to create and scale startups that address social or environmental issues. These entrepreneurs not only strive for financial profitability but also aim to create positive change in society. In the context of climate change and food security, social entrepreneurs play a critical role in developing innovative solutions that are sustainable, inclusive, and scalable.

One key aspect of impact investing and social entrepreneurship is the focus on long-term sustainability. By leveraging their resources, knowledge, and networks, entrepreneurs can develop and implement strategies that not only address immediate challenges but also create lasting solutions. This approach is particularly crucial in the face of climate change, where the need for adaptation and resilience is paramount.

Furthermore, impact investing and social entrepreneurship have the potential to unlock new opportunities for entrepreneurs. As climate change and food security become increasingly urgent issues, there is a growing demand for innovative solutions. Entrepreneurs who can effectively navigate this space have the opportunity to create businesses that not only generate financial returns but also make a meaningful impact on society and the planet.

In conclusion, impact investing and social entrepreneurship offer a powerful avenue for entrepreneurs to tackle climate change and ensure food security. By combining financial viability with positive social and environmental impact, entrepreneurs can drive positive change while also creating sustainable and scalable solutions. As the world faces these pressing challenges, it is imperative for entrepreneurs to embrace impact investing and social entrepreneurship as vital tools in building a better future.

Leveraging Grants and Subsidies for Sustainable Agriculture

In the face of climate change and the urgent need to ensure food security, entrepreneurs play a crucial role in transforming the agricultural sector towards sustainability. One powerful tool at their disposal is the utilization of grants and subsidies. These financial resources can provide the necessary support and incentives for entrepreneurs to develop innovative solutions that address the challenges posed by climate change while ensuring food production for future generations.

Grants and subsidies can be obtained from various sources, including government agencies, non-profit organizations, and private foundations. These funding opportunities are designed to encourage and reward entrepreneurs who are committed to sustainable agricultural practices. By leveraging these financial resources, entrepreneurs can access the necessary capital to implement their ideas and scale their businesses.

One key aspect of these grants and subsidies is their focus on sustainable agriculture. This means that entrepreneurs need to demonstrate how their projects contribute to mitigating climate change, adapting to its impacts, and promoting biodiversity. For instance, an entrepreneur might propose a project that utilizes precision agriculture techniques to minimize water and fertilizer use, reducing greenhouse gas emissions and water pollution.

Additionally, grants and subsidies can support research and development efforts, enabling entrepreneurs to explore and implement cutting-edge technologies. This may include the use of artificial intelligence, remote sensing, or blockchain to optimize resource allocation, improve crop yields, and enhance supply chain transparency. By embracing innovation, entrepreneurs can create more resilient and efficient agricultural systems that are better prepared to withstand the challenges of climate change.

Entrepreneurs must also be aware of the specific requirements and eligibility criteria for each grant or subsidy they seek. It is important to thoroughly research and understand the application process, deadlines, and reporting obligations. Furthermore, entrepreneurs should consider collaborating with research institutions, universities, or other partners to strengthen their proposals and increase their chances of securing funding.

In conclusion, grants and subsidies are invaluable resources for entrepreneurs in the field of sustainable agriculture. They provide the necessary financial support to develop and implement innovative solutions that address climate change and ensure food security. By leveraging these resources effectively, entrepreneurs can make significant contributions to mitigating the impacts of climate change, promoting biodiversity, and building resilient agricultural systems for the future.

Chapter 7: Overcoming Policy and Regulatory Barriers

Advocacy for Climate-Smart Policies and Regulations

Introduction:

In our rapidly changing world, the impact of climate change on food security cannot be ignored. As entrepreneurs, you have a unique opportunity to drive change and make a significant difference in addressing this pressing issue. This subchapter focuses on the importance of advocating for climate-smart policies and regulations, and how you can play a pivotal role in shaping them.

Understanding the Problem: Climate change poses a great threat to global food security. Rising temperatures, erratic weather patterns, and increased frequency of extreme events like droughts and floods have severe implications for agricultural productivity. Entrepreneurs in the niche of climate change must recognize the urgency of this issue and work towards finding sustainable solutions.

The Power of Advocacy: Advocacy is a powerful tool that can create meaningful change at the policy level. By actively engaging with policymakers, stakeholders, and the public, entrepreneurs can influence the development and implementation of climate-smart policies and regulations. Advocacy efforts can help in creating an enabling environment that supports innovation, investment, and adoption of sustainable practices in the agricultural sector.

Key Advocacy Strategies:

1. Building Partnerships: Collaborate with like-minded organizations, communities, and experts to amplify your advocacy efforts. Partnerships can strengthen your voice, provide a platform for knowledge sharing, and widen your reach.

2. Raising Awareness: Educate the public, policymakers, and fellow entrepreneurs about the impacts of climate change on food security. By highlighting the urgency of the issue and showcasing innovative solutions, you can garner support for climate-smart policies and regulations.

3. Engaging in Policy Dialogue: Participate in policy discussions and engage with relevant stakeholders to ensure that climate-smart policies are informed by scientific evidence and practical insights. By providing your entrepreneurial perspective, you can contribute to the development of effective and sustainable solutions.

4. Mobilizing Resources: Advocate for increased financial support and incentives for climate-smart agriculture. This can help entrepreneurs in the climate change niche to overcome barriers and scale up their initiatives, ultimately benefiting food security and the environment.

Conclusion:

Entrepreneurs have a crucial role to play in advocating for climate-smart policies and regulations. By leveraging their expertise, networks, and influence, they can drive positive change towards a more sustainable and resilient food system. Embracing this responsibility is not only an opportunity but a necessity for ensuring a future where food security and climate change mitigation go hand in hand.

Navigating Legal Frameworks for Sustainable Food Ventures

In today's rapidly changing world, entrepreneurs are increasingly recognizing the importance of sustainable food ventures in tackling climate change and ensuring food security. However, venturing into this domain requires more than just innovative ideas and a passion for the environment. Entrepreneurs must also navigate the complex and ever-evolving legal frameworks that govern the food industry. This subchapter aims to provide entrepreneurs in the niche of climate change with a comprehensive understanding of the legal landscape surrounding sustainable food ventures.

The legal frameworks surrounding sustainable food ventures are multi-faceted, encompassing a range of areas such as agriculture, food safety, labeling, waste management, and environmental regulations. Understanding these frameworks is crucial for entrepreneurs to ensure compliance, mitigate risks, and seize growth opportunities.

One of the key aspects entrepreneurs need to navigate is food labeling regulations. As consumers become more conscious of the environmental impact of their food choices, entrepreneurs must accurately communicate the sustainability credentials of their products. This subchapter will delve into labeling requirements, certification programs, and eco-labels, empowering entrepreneurs to effectively communicate their commitment to climate change mitigation.

Another critical aspect is waste management. Sustainable food ventures often prioritize minimizing food waste and optimizing resource utilization. Entrepreneurs must familiarize themselves with regulations regarding waste disposal, recycling, and composting to ensure compliance and implement sustainable waste management practices.

Furthermore, entrepreneurs must understand the legal frameworks surrounding agriculture and farming practices. This includes regulations related to organic farming, genetically modified organisms (GMOs), pesticide use, and land management. Compliance with these regulations not only ensures the sustainability of food production but also helps build consumer trust and market differentiation.

Lastly, this subchapter will explore the legal frameworks related to environmental protection and sustainability. Entrepreneurs must be aware of environmental regulations, carbon footprint reporting, renewable energy incentives, and other climate change mitigation policies. Understanding these frameworks will enable entrepreneurs to leverage available opportunities, such as government grants or tax incentives, to propel their sustainable food ventures forward.

In conclusion, while entrepreneurs in the niche of climate change are driven by their passion for sustainable food ventures, navigating the legal frameworks is essential for success. This subchapter aims to equip entrepreneurs with the knowledge and tools necessary to understand and comply with regulations related to labeling, waste management, agriculture, and environmental protection. By doing so, entrepreneurs can ensure the sustainability of their food ventures while contributing to the broader goal of addressing climate change and ensuring food security for future generations.

Addressing Trade and Market Barriers for Climate-Resilient Food

In the face of climate change, ensuring food security has become a pressing challenge for entrepreneurs in the agriculture and food industries. As the world's population continues to grow, coupled with the increasing impacts of climate change on agricultural productivity, entrepreneurs must find innovative solutions to address trade and market barriers for climate-resilient food.

Trade barriers, both tariff and non-tariff, often hinder the movement of climate-resilient food across borders. Entrepreneurs need to advocate for fair trade policies that promote the exchange of sustainable and climate-resilient food products. This includes reducing trade barriers, streamlining customs procedures, and fostering international cooperation to create a level playing field for entrepreneurs working towards food security in the face of climate change.

Additionally, entrepreneurs must address market barriers that prevent the widespread adoption and consumption of climate-resilient food. Education and awareness campaigns are vital to inform consumers about the benefits of these foods, such as their nutritional value, lower carbon footprint, and resilience to climate shocks. By highlighting the positive impact of climate-resilient food on both human health and the environment, entrepreneurs can create demand and drive market growth.

Furthermore, entrepreneurs should collaborate with governments, research institutions, and non-profit organizations to develop and implement policies that incentivize the production and consumption of climate-resilient food. This could include providing financial support, tax incentives, and research grants to entrepreneurs who are investing in sustainable farming practices, innovative technologies, and climate-resilient crop varieties.

To overcome trade and market barriers, entrepreneurs should also invest in building strong supply chains for climate-resilient food. This involves establishing partnerships with farmers, producers, distributors, and retailers to ensure a consistent and reliable flow of these products to consumers. By strengthening supply chains, entrepreneurs can overcome logistical challenges and ensure the availability of climate-resilient food in both local and global markets.

In conclusion, addressing trade and market barriers for climate-resilient food is crucial for entrepreneurs in the face of climate change. By advocating for fair trade policies, raising awareness among consumers, collaborating with stakeholders, and building robust supply chains, entrepreneurs can play a significant role in promoting the production and consumption of climate-resilient food. Through these efforts, entrepreneurs have the power to tackle climate change and ensure food security for future generations.

Chapter 8: Case Studies: Successful Entrepreneurial Approaches to Climate Change and Food Security

Case Study 1: From Farm to Fork - The Story of a Sustainable Food Startup

Introduction:

In this subchapter, we delve into the inspiring journey of a sustainable food startup that successfully addressed climate change and ensured food security. This case study presents a remarkable example of how entrepreneurs can make a meaningful impact by adopting innovative practices in the face of these pressing challenges.

Background:

The global food system is under tremendous strain due to climate change, population growth, and unsustainable farming practices. Entrepreneurs are uniquely positioned to develop creative solutions that can mitigate these issues and build a sustainable future. This case study serves as a blueprint for entrepreneurs seeking to create a positive impact.

The Birth of the Sustainable Food Startup: Our case study revolves around GreenFields, a sustainable food startup founded by Sarah Thompson. Sarah, a passionate environmentalist, recognized the urgent need for sustainable food production and consumption. She embarked on a mission to create a business that would address climate change while ensuring food security.

Innovative   Farming   Techniques:
Sarah understood the importance of sustainable farming practices and decided to implement a combination of vertical farming and aquaponics. By utilizing vertical space and recirculating water, GreenFields minimized land use and water consumption, while maximizing crop yield. This innovative approach significantly reduced the startup's carbon footprint.

Supply   Chain   Revolution:
GreenFields was committed to creating a closed-loop supply chain. Sarah formed partnerships with local farmers practicing regenerative agriculture, ensuring a constant supply of organic produce. Additionally, she collaborated with local grocery stores and restaurants, establishing a direct farm-to-fork model that eliminated unnecessary intermediaries. This approach not only reduced food waste but also decreased transportation emissions.

Educating   Consumers:
Recognizing the importance of consumer education, GreenFields launched an awareness campaign highlighting the positive impact of sustainable food choices. They organized workshops, cooking classes, and community events to educate people about the benefits of locally sourced and sustainably produced food. This not only increased consumer demand for their products but also empowered individuals to make more informed choices.

Conclusion:

The success story of GreenFields showcases how entrepreneurs can play a pivotal role in addressing climate change and ensuring food security. By adopting innovative farming techniques, revolutionizing supply chains, and educating consumers, Sarah Thompson and her team created a sustainable food startup that serves as an inspiration to a new generation of entrepreneurs. This case study highlights the endless possibilities for entrepreneurs to make a positive impact on both the environment and the global food system.

Case Study 2: Empowering Women Farmers in Climate-Resilient Agriculture

Introduction:

In this case study, we delve into the inspiring story of how empowering women farmers can not only combat the challenges of climate change but also ensure food security. As entrepreneurs committed to addressing climate change, it is crucial to recognize the vital role women play in sustainable agriculture and the potential for transformative impact.

The   Context:

Climate change poses significant threats to agriculture, including unpredictable weather patterns, water scarcity, and increased pest and disease outbreaks. These challenges disproportionately affect vulnerable communities, exacerbating food insecurity and poverty. However, women, who constitute a substantial portion of the agricultural workforce, can be powerful agents of change when equipped with the right resources and opportunities.

The   Case   Study:

This case study focuses on a project implemented in a rural community in a developing country. The project aimed to empower women farmers by providing them with climate-resilient agricultural practices, access to resources, and training programs. The initiative also aimed to enhance their participation in decision-making processes, ensuring their voices were heard.

1.   Building   Climate   Resilience:

The project began by introducing climate-smart farming

techniques such as crop diversification, agroforestry, and water conservation methods. By adopting these practices, women farmers were better equipped to adapt to changing weather patterns, conserve natural resources, and reduce greenhouse gas emissions.

2. Access to Resources and Training: Recognizing the importance of access to resources, the project provided women farmers with improved seeds, fertilizers, and tools. Additionally, training programs were conducted to enhance their knowledge and skills in sustainable farming practices, financial management, and entrepreneurship. These initiatives enabled women to increase their productivity, incomes, and overall well-being.

3. Women's Empowerment and Participation: The project also focused on empowering women by promoting gender equality and enhancing their participation in decision-making processes. Through targeted workshops and capacity-building programs, women farmers gained confidence, leadership skills, and the ability to advocate for their needs and rights.

Results and Impact: The project yielded remarkable results, demonstrating the potential of empowering women farmers in climate-resilient agriculture. Women's incomes and food security improved significantly, contributing to poverty reduction in the community. Moreover, the adoption of sustainable farming practices led to enhanced environmental stewardship and increased resilience to climate change.

Conclusion:

This case study highlights the transformative power of empowering women farmers in climate-resilient agriculture. As entrepreneurs, we have a unique opportunity to contribute to sustainable development by investing in initiatives that promote gender equality, empower women, and tackle climate change. By recognizing the potential of women as change agents in agriculture, we can work towards a more secure and resilient future for all.

Case Study 3: Scaling Up Sustainable Aquaculture for Food Security

Introduction:

In this chapter, we will explore a compelling case study that highlights the potential of sustainable aquaculture in addressing the pressing challenges of food security and climate change. As entrepreneurs, you have the opportunity to make a significant impact by embracing innovative solutions that not only mitigate the effects of climate change but also ensure a steady supply of nutritious food for future generations.

Case   Study   Overview:
Aquaculture, the farming of aquatic organisms such as fish, shellfish, and aquatic plants, has gained prominence as a sustainable alternative to traditional fishing practices. This case study focuses on the successful scaling up of sustainable aquaculture initiatives to tackle food security challenges while reducing environmental impacts.

Background:

With a rapidly growing global population, ensuring food security has become a paramount concern. Climate change exacerbates this challenge, as it disrupts traditional farming practices and negatively impacts natural ecosystems. Sustainable aquaculture offers an alternative that can provide a consistent supply of protein-rich food, while also mitigating the environmental consequences associated with overfishing.

Key   Findings:
1. Sustainable aquaculture practices, such as integrated multi-trophic aquaculture and recirculating aquaculture systems, have proven effective in minimizing environmental impacts, including water pollution and habitat destruction.
2. By adopting responsible feed sourcing and reducing reliance on wild-caught fish for fishmeal, entrepreneurs can support the sustainable growth of aquaculture.
3. Innovations in aquaculture technology, such as automated monitoring systems and precision feeding, enhance productivity while minimizing resource inputs and waste generation.
4. Engaging local communities and promoting social responsibility are crucial for successful scaling up of sustainable aquaculture initiatives.
5. Collaboration between entrepreneurs, governments, and research institutions is essential to develop policy frameworks, share best practices, and foster innovation in the aquaculture sector.

Opportunities   for   Entrepreneurs:
1. Investing in sustainable aquaculture ventures offers the potential for profitable returns while addressing food security and climate change concerns.
2. Developing and marketing sustainable aquaculture technologies, such as efficient recirculating systems or eco-friendly feed alternatives, can open new business opportunities.
3. Building partnerships with local communities, scientists, and policymakers can facilitate the successful implementation of

sustainable aquaculture projects.
4. Educating consumers about the benefits of sustainably sourced aquaculture products can create a demand-driven market and lead to wider adoption.

Conclusion:

Scaling up sustainable aquaculture holds immense potential for entrepreneurs to tackle the intertwined challenges of climate change and food security. By embracing innovative practices, technologies, and collaborations, entrepreneurs can contribute to a more sustainable and resilient global food system. Let this case study inspire you to explore the vast opportunities that sustainable aquaculture offers and be part of the solution in feeding the future.

Chapter 9: The Future of Entrepreneurship in Climate-Smart Food Systems

Emerging Trends and Opportunities

In the face of climate change and the increasing global concern for food security, entrepreneurs have an unparalleled opportunity to create innovative solutions that can tackle these challenges head-on. This subchapter explores the emerging trends and opportunities that lie at the intersection of entrepreneurship, climate change, and food security.

1. Sustainable Agriculture: As the world population continues to grow, the demand for food is expected to increase significantly. Entrepreneurs who embrace sustainable agriculture practices, such as precision farming, vertical farming, and hydroponics, can not only ensure higher crop yields but also reduce the environmental impact of traditional farming methods.

2. Plant-Based Alternatives: The shift towards plant-based diets is gaining momentum, driven by environmental concerns and health considerations. Entrepreneurs can capitalize on this trend by developing innovative plant-based alternatives to traditional animal-based products. From meat substitutes to dairy-free alternatives, the opportunities are vast for entrepreneurs to create sustainable and nutritious food options.

3. Food Waste Reduction: Food waste is a significant contributor to greenhouse gas emissions. Entrepreneurs can play a vital role in reducing food waste by developing technologies and business

models that enable efficient food distribution, preservation, and waste management. From food surplus redistribution platforms to composting solutions, there are numerous opportunities for entrepreneurs to make a positive impact.

4. Agri-Tech and Data Analytics: The digital revolution has paved the way for advancements in agriculture technology and data analytics. Entrepreneurs can leverage these technologies to optimize farming practices, enhance crop monitoring, and improve supply chain efficiency. By harnessing the power of data, entrepreneurs can make informed decisions that not only increase productivity but also mitigate the impact of climate change on food security.

5. Climate-Resilient Farming: As climate change brings about more frequent extreme weather events, entrepreneurs can explore opportunities to develop and promote climate-resilient farming practices. This includes drought-tolerant and climate-adaptive crops, as well as innovative irrigation and water conservation techniques. By embracing climate-resilient farming, entrepreneurs can ensure the long-term sustainability of agriculture in the face of a changing climate.

In conclusion, the emerging trends and opportunities highlighted in this subchapter present a compelling case for entrepreneurs to step up and make a difference in addressing climate change and ensuring food security. By embracing sustainable agriculture practices, developing plant-based alternatives, reducing food waste, leveraging agri-tech and data analytics, and promoting climate-resilient farming, entrepreneurs have the potential to create innovative solutions that will shape the future of food production and consumption. The time is ripe for entrepreneurial action – the future of our planet and the well-being of future generations depend on it.

The Role of Technology in Transforming Food Systems

In today's rapidly changing world, where climate change poses a significant threat to our food security, entrepreneurs have a crucial role to play in finding innovative solutions. The integration of technology in transforming food systems has become an imperative to ensure a sustainable future. This subchapter explores the pivotal role that technology plays in addressing the challenges of climate change and achieving food security.

Technology has the power to revolutionize every aspect of our food systems, from production to distribution and consumption. It enables us to optimize resource utilization, reduce waste, and increase productivity in the face of climate change-induced challenges. With the right technological advancements, entrepreneurs can create a more resilient and sustainable food system.

One of the key areas where technology can make a significant impact is in precision agriculture. By leveraging tools such as sensors, drones, and satellite imagery, entrepreneurs can monitor and manage crops more efficiently. This allows for precise application of resources like water, fertilizers, and pesticides, reducing waste and enhancing productivity. Additionally, the use of artificial intelligence and machine learning can enable predictive analytics, helping farmers make informed decisions and mitigate climate-related risks.

Moreover, technology can enable the development of alternative and sustainable food production methods. Vertical farming, hydroponics, and aquaponics are examples of innovative approaches that use technology to grow food in controlled environments, reducing the reliance on traditional agricultural practices. Entrepreneurs can leverage these technologies to create scalable and efficient solutions that can be implemented in both urban and rural settings, ensuring year-round food production and reducing the environmental impact.

In the realm of food distribution and logistics, technology can enhance efficiency and reduce waste. The use of blockchain, for instance, can provide transparency and traceability throughout the supply chain, ensuring food safety and quality. Furthermore, innovative packaging and preservation technologies can extend the shelf life of perishable items, reducing post-harvest losses and improving access to fresh produce.

In conclusion, the role of technology in transforming food systems cannot be overstated. Entrepreneurs have a unique opportunity to harness the power of technology to tackle the challenges of climate change and ensure food security. By embracing technological advancements in precision agriculture, sustainable food production, and efficient distribution systems, entrepreneurs can drive positive change and create a more resilient and sustainable future. The time to act is now, and through innovation and collaboration, we can feed the future while mitigating the impact of climate change.

The Importance of Continuous Innovation and Adaptation

In today's rapidly changing world, the ability to continuously innovate and adapt is crucial for entrepreneurs, especially those operating in the niche of climate change. In the face of pressing challenges such as global warming and food insecurity, it is imperative that entrepreneurs adopt a forward-thinking mindset and embrace continuous innovation to address these issues effectively.

One of the key reasons why continuous innovation is paramount is the ever-evolving nature of climate change. As the planet undergoes significant shifts in weather patterns and experiences the consequences of human activities, entrepreneurs must adapt their strategies and offerings to stay relevant. By continuously innovating, entrepreneurs can develop sustainable solutions that mitigate the impacts of climate change and contribute to long-term food security.

Furthermore, continuous innovation allows entrepreneurs to seize new opportunities and stay ahead of the competition. As awareness of climate change and its consequences grows, consumers are becoming more conscious of their choices. They are actively seeking sustainable and eco-friendly alternatives. By constantly innovating, entrepreneurs can create products and services that cater to this growing demand, capturing a larger market share and establishing themselves as leaders in the fight against climate change.

Adaptation is another crucial aspect that entrepreneurs need to embrace. The ability to adapt to changing circumstances is essential when addressing climate change. Entrepreneurs must be flexible and willing to modify their strategies and business models as new challenges arise. This includes embracing emerging technologies, exploring alternative supply chains, and adopting sustainable practices that reduce their environmental impact.

Moreover, continuous innovation and adaptation go hand in hand with resilience. Climate change brings about unpredictable disruptions, such as extreme weather events and resource scarcity, that can severely impact food production and distribution. By continuously innovating and adapting, entrepreneurs can build resilience into their operations, enabling them to navigate these challenges effectively and ensure food security for future generations.

In conclusion, the importance of continuous innovation and adaptation cannot be overstated for entrepreneurs operating in the niche of climate change. By embracing a forward-thinking mindset and continuously innovating, entrepreneurs can develop sustainable solutions, capture market share, and contribute to mitigating the impacts of climate change. Additionally, adaptation allows entrepreneurs to build resilience into their operations, ensuring food security amidst the uncertainties brought about by a changing climate. It is through continuous innovation and adaptation that entrepreneurs can truly make a difference in tackling climate change and ensuring a sustainable future for all.

Chapter 10: Conclusion: Empowering Entrepreneurs to Feed the Future

Recap of Key Insights and Learnings

As entrepreneurs in the face of climate change, it is crucial that we understand the significance of our role in ensuring food security for the future. In this subchapter, we will recap some of the key insights and learnings from our book, "Feeding the Future: How Entrepreneurs Can Tackle Climate Change and Ensure Food Security," specifically tailored to entrepreneurs like yourselves who are passionate about tackling climate change and working towards a sustainable future.

One of the fundamental insights we have explored is the link between climate change and food security. Climate change poses a significant threat to our global food systems, with extreme weather patterns, rising temperatures, and changing precipitation patterns impacting agricultural productivity. Entrepreneurs must recognize this connection and take proactive measures to build resilient and sustainable food systems.

We have also emphasized the importance of adopting innovative and sustainable agricultural practices. Entrepreneurs can play a transformative role by investing in technologies like precision agriculture, vertical farming, and hydroponics. These technologies not only increase productivity but also reduce resource consumption and minimize environmental impact.

Furthermore, we have highlighted the significance of collaboration and partnerships. Entrepreneurs need to work together with farmers, policymakers, scientists, and other stakeholders to develop comprehensive solutions. By fostering collaborations, sharing knowledge, and leveraging collective expertise, we can create a more resilient and sustainable food ecosystem.

Another key learning we have explored is the potential of alternative protein sources. As the global demand for protein continues to rise, entrepreneurs can explore and invest in plant-based proteins, cultured meat, and insect proteins. These alternatives not only have a lower environmental footprint but also offer new business opportunities and address the challenges of food security.

Lastly, we have emphasized the importance of consumer education and behavior change. Entrepreneurs can play a pivotal role in raising awareness about the environmental impact of food choices and promoting sustainable diets. By influencing consumer preferences and habits, we can drive demand for sustainable food products and contribute to a more climate-friendly food system.

In conclusion, as entrepreneurs addressing climate change, we must understand the critical role we play in ensuring food security for the future. By adopting innovative practices, fostering collaborations, exploring alternative protein sources, and promoting consumer education, we can make a significant impact. Together, we have the power to create a sustainable future that not only addresses climate change but also ensures food security for generations to come.

Call to Action for Entrepreneurs

As entrepreneurs, we hold immense power to shape the future of our planet and address the pressing challenges of climate change and food security. In the face of these global crises, it is our duty to step up and take action. This subchapter aims to ignite a sense of urgency and purpose within entrepreneurs, urging them to seize the opportunity and become catalysts for change.

The world is at a critical juncture, with climate change threatening the very foundations of our existence. Rising temperatures, extreme weather events, and diminishing resources are all indicators of an impending crisis. Simultaneously, the issue of food security looms large, with a growing global population and increasing pressure on our food systems.

Entrepreneurs must recognize that climate change and food security are not isolated problems but interconnected challenges. By addressing one, we can positively impact the other. It is through innovation, creativity, and collaboration that we can create sustainable solutions that ensure a prosperous future for all.

The call to action for entrepreneurs is multifaceted. Firstly, we must embrace the potential of technology and innovation to revolutionize our food systems. From precision agriculture and vertical farming to alternative protein sources and sustainable packaging, there is a vast array of opportunities waiting to be explored. By harnessing the power of entrepreneurship, we can develop new business models that prioritize environmental sustainability and social responsibility.

Secondly, entrepreneurs must foster partnerships and collaborations. Tackling climate change and ensuring food security requires collective action. By joining forces with like-minded individuals, organizations, and governments, we can amplify our impact and drive systemic change. Collaboration enables us to share knowledge, resources, and expertise, accelerating the development and implementation of sustainable solutions.

Thirdly, entrepreneurs must be advocates for policy change. The transformation we seek cannot be achieved in isolation; it requires supportive policies and regulations. By engaging with policymakers, raising awareness about the urgency of the issues, and advocating for sustainable practices, entrepreneurs can influence the creation of an enabling environment that incentivizes and rewards sustainable entrepreneurship.

Lastly, entrepreneurs must lead by example. We must adopt sustainable practices within our own businesses, embracing circular economy principles, reducing waste, and minimizing our carbon footprint. By demonstrating the feasibility and profitability of sustainable business models, we can inspire others to follow suit.

In conclusion, the call to action for entrepreneurs is clear. We have the power, the creativity, and the responsibility to tackle climate change and ensure food security. Let us seize this opportunity, embrace innovation, collaborate, advocate for change, and lead by example. Together, we can create a sustainable future for generations to come.

Inspiring a Sustainable and Food-Secure Future

In today's world, the dual challenges of climate change and food security have become increasingly pressing. As entrepreneurs, we have a unique opportunity to make a positive impact and contribute to a sustainable and food-secure future. By recognizing the connection between these two issues and harnessing our innovative spirit, we can create transformative solutions that address the urgent needs of our planet and its growing population.

The first step towards inspiring a sustainable and food-secure future is understanding the linkages between climate change and food production. Climate change poses a severe threat to agriculture, with rising temperatures, changing precipitation patterns, and extreme weather events disrupting crop yields and livestock productivity. Additionally, the agricultural sector itself contributes to greenhouse gas emissions, further exacerbating the climate crisis. By recognizing these interdependencies, we can develop strategies that tackle both challenges simultaneously.

Entrepreneurs in the field of climate change have already made significant strides in developing clean energy solutions, sustainable transportation systems, and carbon capture technologies. However, it is crucial to extend our focus to the food system, which is responsible for a significant portion of global greenhouse gas emissions. By investing in regenerative agriculture techniques, promoting sustainable farming practices, and developing innovative technologies for food production, we can significantly reduce the carbon footprint of our food system.

Furthermore, ensuring food security goes beyond just addressing climate change. It involves improving access to nutritious and affordable food for all, especially in vulnerable communities. Entrepreneurs can play a crucial role in developing innovative business models that increase the efficiency of food distribution, reduce food waste, and promote sustainable consumption patterns. By partnering with local communities, governments, and NGOs, we can create inclusive solutions that uplift small-scale farmers and empower them to produce more food sustainably.

To inspire a sustainable and food-secure future, collaboration and knowledge-sharing are essential. Entrepreneurs can form alliances with other like-minded individuals and organizations, leveraging their collective expertise and resources. By joining forces, we can accelerate the development and adoption of sustainable agricultural practices, renewable energy solutions, and climate-smart technologies.

Conclusion,

as entrepreneurs, we have the power to shape a sustainable and food-secure future. By recognizing the interconnections between climate change and food security, we can develop innovative solutions that address both challenges simultaneously. With collaboration and a shared commitment to a sustainable future, we can pave the way for a more resilient and thriving planet for generations to come. Together, let us seize this opportunity to make a lasting impact and ensure a brighter future for all.

Appearance

Although bats exist in a vast range of shapes and sizes, they all have a few basic traits in common. They have two wings consisting of skin stretched between long fingers and are mammals with fur. Here are some characteristics of bat look in general:

Bats have small heads and long, slender bodies.

A thin skin membrane known as the patagium covers their wings, which have been modified for flight.

In some species, the patagium can also cover the animal's legs and tail in addition to its fingers and torso.

Bats can utilize their enormous, frequently pointed ears for echolocation to help them travel and find prey.

They have lengthy tongues for slurping nectar or fruit juice, and keen teeth for biting and breaking their meal.